Written by J.R. Becker & illustrated by Max Rambaldi

An evolution story of mass ~~proportions~~ extinctions

In memory of Angie Warren

Beloved wife, biologist, dreamer, and inspiration

"The earth does not belong to man; man belongs to the earth. All things are connected like the blood that unites one family. Man did not weave the web of life, he is merely a strand in it. Whatever he does to the web, he does to himself."

Chief Seattle

Annabelle & Aiden
breathed in the fresh air.
They drank from a brook.
The sun warmed their hair.
“Our world really seems
like it fits us so well!
Was it custom-made
as a place we could dwell?”

"Of course!" Aiden said,
"All this air, land and sea.
I feel it, Annabelle:
this world
was made for me!"

"I've heard those words before..."
Tardigrade Tom waved his cane,
"But it's not so simple, kids.
Just allow me to explain..."

Once upon a time,
the first living things
had many different babies,
who lived like tiny kings.

"We have all the stuff we need,"
they had started to discuss,
"in a sea that's danger-free.
Oh, this world was made for us!"

THE GREAT OXYGENATION EVENT
(2.4 billion yrs ago)

Some microbes evolved a more-efficient photosynthesis that produced oxygen as a waste-product, which was toxic to the others, causing almost all life on Earth to go extinct.

Until they had a baby
a bit different from them,
who burped out tons of stuff
new to them, called oxygen.

They couldn't breathe this stuff
that they'd never seen before.
Or live in this new world
and so, soon they were no more.

The “brother” she absorbed,
she used as an internal photosynthesis factory;
as she became the ancestor of all plant cells,
the “brother” inside turned into what we now
call chloroplasts.

She could breathe the oxygen
in a way that no one had!
This was a random surprise
for which she was very glad.

Except one random one,
who just happened to have ate
her oxygen-burping brother,
which then caused her to mutate...

The oxygen-rich waters
were perfect for them to breathe.
So they smiled and exclaimed,
"This world was made for me!"

She had babies
who had babies,
who had babies more complex:
worms and fish who filled the sea,
each one stranger than the next.

Until roots of nearby plants pushed loose dirt into the sea. Even worse, the dirt transformed into clouds of thick algae!

ORDOVICIAN-SILURIAN & DEVONIAN EXTINCTIONS

370-445 million yrs ago

Which sucked up the oxygen,
and crowded the ocean floor.
With no space or air to breathe,
fish could not live any more.

Except one lucky fish
with fins a bit like hands,
whose strange mutation let
him climb up onto land.

He had babies like himself,
with webbed feet to walk and swim.
They had babies of their own,
reptile and amphibian.

These muddy shores were perfect
for their newly-formed webbed feet.
"It's so obvious," they said,
"This world was made for me!"

Cacops

Seymouria

Ctenospondylus

Hylonomus

CARBONIFEROUS PERIOD

(350–300 million yrs ago)

America

Pareiasaurus

Dimetron

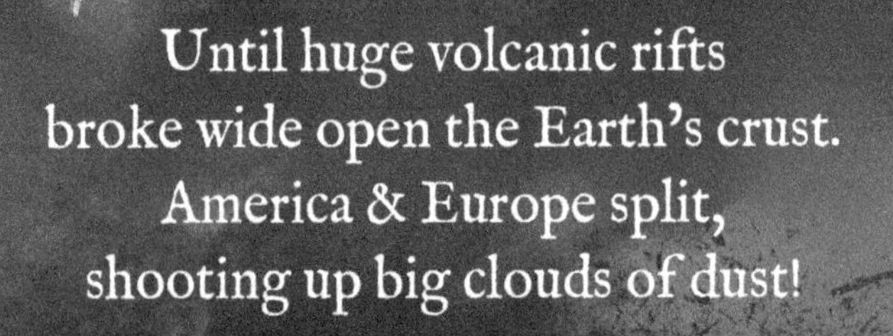

Until huge volcanic rifts
broke wide open the Earth's crust.
America & Europe split,
shooting up big clouds of dust!

PERMIAN, TRIASSIC & JURASSIC EXTINCTION

200-260
million yrs ago

Volcanoes then erupted
through the Atlantic sea.
Clouds of smoke, gas and fire
made it very hard to breathe.

The reptiles perished.
Except one of them that had
strangely different kinds of lungs
for which she was very glad.

They helped her breathe real well.
This small dino-like reptile,
she had babies who had babies
who grew **huge** after a while.

They said, "This world's air is so
perfect for our lungs to breathe.
And there's few others around.
So this world was made for me!"

Until one day they saw a
tiny dot up in the sky.
It grew bigger and bigger,
'till it wasn't very high.

It was coming straight for them!
An asteroid! A fireball!
The same size of Mt. Everest,
and they couldn't stop its fall.

It smashed into Mexico.
Plants & trees burst into flame.
Then waves one thousand feet high,
washed the rest of them away.

Clouds of ash blocked out the sun.
The world froze under snow.
All the animals then died.
Since they had nowhere to go.

Except one tiny creature
like a squirrel, furry and brown,
who just happened to be small
and could burrow underground.

With feet also good for climbing,
and for swinging from the trees.
With dinos gone, things were great,
as she lived on nuts and seeds.

She had babies that grew bigger
from the plants they found to eat.
Lemurs, lorises, and monkeys,
all with perfect climbing feet.

New soil grew thick forests.
Gibbons swung from tree to tree.
Primates looked around and said,
"Oh, this world was made for me!"

Until India then crashed
into land we call Asia,
and made tall mountain ranges
called the Himalayas.

Which blocked rain from reaching
the primates' forest home,
which became empty fields
they were not able to roam.

Except for one of them,
who was born with something strange:
big flat toes that helped her walk
across bare fields, for a change.

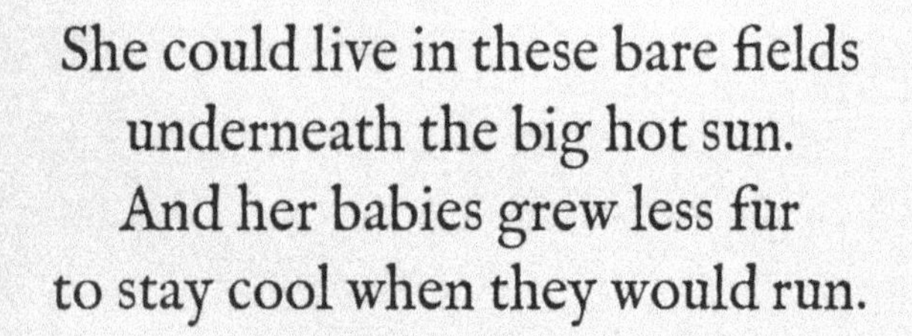

She could live in these bare fields
underneath the big hot sun.
And her babies grew less fur
to stay cool when they would run.

They got smarter and made clothing
and spread out to different lands.
Cut down trees, and built cities,
and got food from cows and plants.

They breathed air cuz that first thing
could breathe different from the rest.
They had legs from that first fish
that climbed up to take a breath.

And toes to walk, just because
India's path had blocked the rain.
They forgot that all the rest
swung from trees and died away.

They forgot about all them:
all the ones who came and went.
All the flawed, extinct designs.
Yes, the ninety-nine percent.
Uhm, Aiden, I think it's getting warmer..
They forgot they just showed up.
In no time, they all would say,
"Everything was made for me,
and it's always been this way."
They just see how things are now.
As they dominate proudly.
Puffing out their chests, they boast...
"This World Was Made for ME!"

Skeptisaurus then explained,
"It just looks that way right now:
that this world was built for us,
but it's the other way around!

"You're thinking like a puddle
who sees the hole he's in,
and excitedly declares
how its shape was made for him!

Our world is the way it is.
And we managed to adapt.
If our world had turned out differently,
we would have too, in fact.

If our air was grape juice and
it was always dark outside,
maybe we'd have fins to swim
and four super-powered eyes!

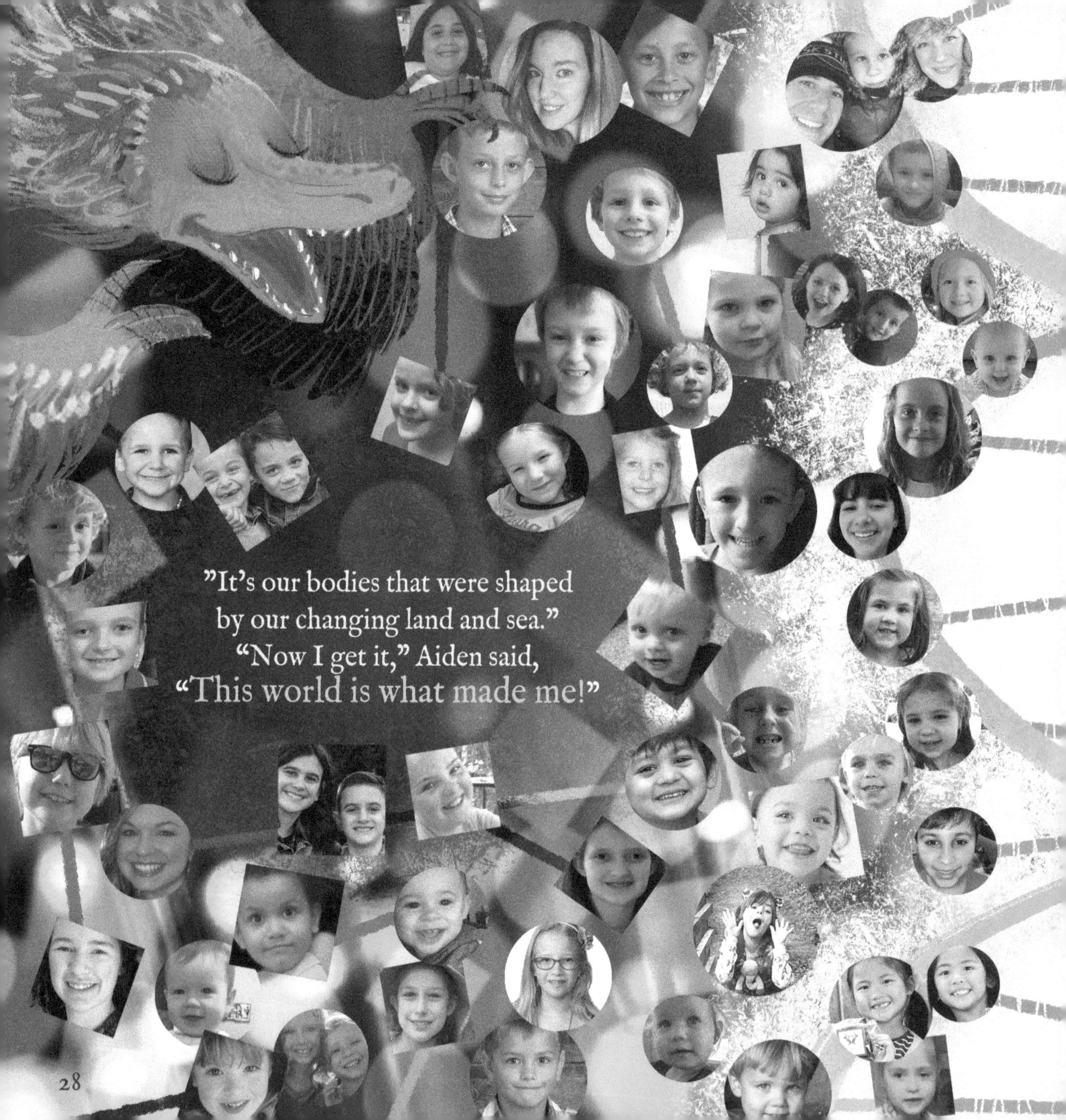

"It's our bodies that were shaped
by our changing land and sea."
"Now I get it," Aiden said,
"This world is what made me!"

Our world is what shaped us.
It's allowed us to visit.
It does not belong to us.
Rather, we belong to it.

We're a strand of a larger
web of life, and all it does.
And if we are good to it,
then it will be good to us.

Author's note

Does our world seem to fit our bodies so well? Of course it does! (Indeed, it would be quite a miracle if it didn't.) That's because we evolved here - not somewhere else. Our changing world has killed out over 99% of life, and we are the refined survivors, whose random genetic mutations help us make it through. It's not that we were "given" lungs to breathe, or eyes to see, or hearts to pump: just that those that *didn't* have those random mutations, which is most, died out. Like most ideas in a company's brain-storming session: only a few make it through and are continuously refined. Essentially, every single aspect of our bodies was once a tiny random genetic mutation (the vast majority of which are useless or even harmful), which was then found to be useful in a particular time and place, and was thus naturally selected-for, passed on, and continuously refined over countless generations. Those who weren't lucky enough to end up with those random mutations - which are most - didn't survive, and were discarded as shoddy, unintelligent designs. Most didn't make it. We're the survivors. The few models that made it through. So of course our bodies seem "built" for this world, and thus, from our absurdly limited perspective, this world can seem designed "for" us. If we had evolved on a different planet, or in a different environment, different mutations would have been selected, and our bodies would have been suited for that environment, instead of this one. Perhaps somewhere, in another universe, where the atmosphere is full of helium, is some other species amazed they could breathe it perfectly, foolishly claiming, "This world was made for me!" Why does this misconception matter? It effects how we see our world, and our fragile place within it. To quote someone more eloquent than myself...

"This is rather as if you imagine a puddle waking up one morning and thinking, 'This is an interesting world I find myself in – an interesting hole I find myself in – fits me rather neatly, doesn't it? In fact it fits me staggeringly well, must have been made to have me in it!' This is such a powerful idea that as the sun rises in the sky and the air heats up and as, gradually, the puddle gets smaller and smaller, frantically hanging on to the notion that everything's going to be alright, because this world was meant to have him in it, was built to have him in it; so the moment he disappears catches him rather by surprise. I think this may be something we need to be on the watch out for."

Douglas Adams in The Salmon of Doubt

Collect Our Other Titles at **AnnabelleAndAiden.com**

"A beautiful, whimsical, and **deeply important book** for kids of all ages!"
- Cara Santa Maria, host of *Talk Nerdy Podcast*, co-author of *The Skeptic's Guide To The Universe*

"A great book. **Very smart. And kind.**"
- Penn Jillette, author of *God, No!*, comedian, magician of *Penn & Teller*

"Does an exemplary job of **explaining the origins of life** as we know it **to very young minds.** Heck, it could probably even teach a thing or two to their folks." - *Bill Nye Film*

"What a stunning & **refreshing set of books**! The exact words & tone to reach into a child's heart and brain and bring out such bliss & beauty. A wonderful addition fo our family library!"
- Mayim Balik, PhD, actress, author, and neuroscientist

"**Beautifully illustrated** books for children that opens their minds and hearts to the wonders of science. Any child who reads it will find themselves **mesmerized, enlightened, and smiling.**"
- Chip Walter, National Geographic Fellow and author of *Last Ape Standing: The Seven-Million-Year Story of How and Why We Survived*

"The most deeply moving account I've ever come across on the ultimate question of life, death, and meaning. **In only 200 words** this beautiful story **captures what 2000 years of wisdom writing could not**. Read it. Then read it again. And then get back to living your life in a way that matters because now you know how."
- Michael Shermer, *New York Times* Bestselling Author of *The Believing Brain*, monthly columnist at *Scientific American*

"Gorgeous and **amazing!**" - Dale McGowan, author of *Parenting Beyond Belief*

Paperback: 978-1-7334752-1-1
Published in the United States by Imaginarium Press, LLC.

CPSIA information can be obtained
at www.ICGtesting.com
Printed in the USA
BVHW022245150820
586545BV00019B/920